I0813136

CHEERS FOR CAREERS!

I WANT TO BE A PLUMBER

by Julie Murray

Cody Koala
An Imprint of Pop!
popbooksonline.com

Hello! My name is Cody Koala

This book is filled with videos, puzzles, games, and more! Scan the QR codes* while you read, or visit the website below to make this book pop.

popbooksonline.com/plumber

*Scanning QR codes requires a web-enabled smart device with a QR code reader app and a camera.

abdobooks.com
Published by Pop!, a division of ABDO, PO Box 398166, Minneapolis, Minnesota 55439.

Printed in the United States of America, North Mankato, Minnesota.
052024
082024

Cover Photo: Shutterstock Images
Interior Photos: Shutterstock Images; Getty Images
Editor: Elizabeth Andrews and Grace Hansen
Series Designer: Colleen McLaren

Library of Congress Control Number: 2023947423

Publisher's Cataloging-in-Publication Data
Names: Murray, Julie, author.
Title: I want to be a plumber / by Julie Murray
Description: Minneapolis, Minnesota : Pop!, 2025 | Series: Cheers for careers! | Includes online resources and index
Identifiers: ISBN 9781098246068 (lib. bdg.) | ISBN 9781098246624 (ebook)
Subjects: LCSH: Plumbing--Juvenile literature. | House drainage--Juvenile literature. | Plumbing industry--Juvenile literature. | Servicing trades--Juvenile literature. | Occupations--Juvenile literature.
Classification: DDC 696.1--dc23

Table of Contents

Chapter 1

Water Flowing

Plumbers work with water. They **install** and repair plumbing systems. They work in homes, offices, stores, and schools. Plumbers work anywhere there is water flowing in pipes!

Watch a video here!

Chapter 2

Skills Needed

Plumbers work with homeowners, business owners, and builders. They need to have good communication skills. They must listen to a problem and share how they plan to fix it.

Plumbers are in high demand in the United States.

drain auger
pipe cutter
torch
plunger
pipe wrench
APP GAS
DANGER
16 oz./453.6 gm.

Plumbers perform tasks with their hands. They use tools such as wrenches, pipe cutters, and torches. They sometimes have to work in small spaces.

Plumbers should have good problem solving skills. Pipes can leak, break, or become clogged. Plumbers must figure out the issue and come up with a plan to fix it.

Chapter 3

How to Become a Plumber

A good first step to becoming a plumber is taking classes at a **technical college**. Then one takes part in an **apprenticeship**. This hands-on work provides a solid base for one's career.

An apprenticeship lasts four to five years.

Explore links here!

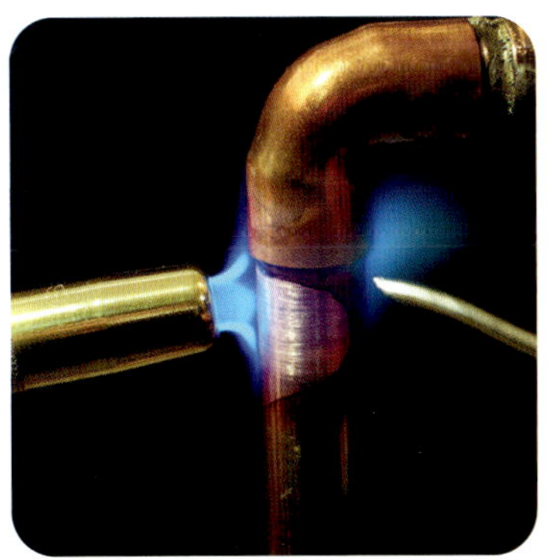

Neyer Plumbing
Neyer Plumbing
KEN NEYER

Next, there is a plumbing exam to become a **journeyman**. Years of working at this level are required before one can move to the next level.

To become a master plumber, a person needs even more experience. There is also an exam. Master plumbers can own their own plumbing business.

It can take up to ten years to become a master plumber.

Chapter 4

Different Jobs

Some plumbers work in **construction**. They create plans for plumbing in a new space. They **install** the pipes and **equipment**. They make sure all water systems work properly.

There are nearly 500,000 **licensed** plumbers in the United States.
Complete an activity here!

Others work on existing plumbing. They find and fix issues. They are problem solvers that help people. Being a plumber is an important and a rewarding career!

Making Connections

Text-to-Self

Has a plumber ever had to fix something at your house? If so, what was it?

Text-to-Text

Can you think of another book that talks about plumbers or other people with important jobs?

Text-to-World

Plumbers work with many tools. In what other careers do people work with tools?

Glossary

apprenticeship – an arrangement in which someone learns a trade under an expert.

construction – the building of something, typically a large structure.

equipment – supplies necessary for a service or action.

install – to put into position and make ready for use.

journeyman – a worker who has learned a trade and works for another person.

licensed – having an official license that gives legal permission to do something.

technical college – a school that educates and prepares students for a specific trade or career.

Index

Online Resources

popbooksonline.com

Thanks for reading this Cody Koala book!

This book is filled with videos, puzzles, games, and more! Scan the QR codes* while you read, or visit the website below to make this book pop.

*Scanning QR codes requires a web-enabled smart device with a QR code reader app and a camera.